YOUR KNOWLEDGE HAS VALUE

- We will publish your bachelor's and
  master's thesis, essays and papers

- Your own eBook and book -
  sold worldwide in all relevant shops

- Earn money with each sale

Upload your text at www.GRIN.com
and publish for free

Ataliba Miguel

Corrosiveness of Different Types of Water

GRIN Verlag

Bibliografische Information der Deutschen Nationalbibliothek:

Die Deutsche Bibliothek verzeichnet diese Publikation in der Deutschen National-
bibliografie; detaillierte bibliografische Daten sind im Internet über http://dnb.d-
nb.de/ abrufbar.

Imprint:

Copyright © 2013 GRIN Verlag GmbH
Druck und Bindung: Books on Demand GmbH, Norderstedt Germany
ISBN: 978-3-656-60806-6

The Robert Gordon University

Corrosiveness of various types of water

Materials & Corrosion Science
Technical Report

Ataliba Miguel

ABSTRACT

The corrosiveness of three different types of water was analysed through an experiment carried out with iron nails immersed in a bath for a period of 25 days. The types of water used during the experiment were respectively synthetic seawater, mineral water and sparkling water. The iron nails immersed in a bath of both synthetic seawater and mineral water suffered a corrosive attack characterized by the formation of reddish brown flakes which have adhered the surface of the iron nails. Whilst, on the other experiment, the iron nail immersed in the sparkling water suffered a very minimal corrosive attack owing to the formation of a protective layer which prevented the underlying steel from further dissolution.

List of Figures

List of Tables

List of graphs

1. Introduction

Corrosion of iron nails was analyzed through an observation experiment carried out in three different types of water to compare the difference in corrosiveness of some commercial drinking waters (mineral water, i.e. no gas, and sparkling water, i.e. with gas) and some synthetic seawater. The length period of observations was approximately 25 days.

Ph strip papers were used to measure the level of ph in the three different types of water. The frequency of ph measurement was every 3 to 4 days. Room temperature was logged and recorded every time ph measurement was taken. Details of the logbook table are available in the annex I.

1.1.   Objectives

The aim of this report is to describe the corrosion of iron nails under stagnant water conditions. Observations, deductions and conclusions will be further detailed throughout the report.

1.2.   Methodology

The methodology adopted to carry out this experiment, was through a combination of visual observations, supplemented by further research available on several studies carried out on corrosion phenomena.

2. Experiment layout

For this experiment 3 iron nails of 110mm x 2.5mm were polished with dry fine emery paper until nice and shiny. Two commercial drinking water (mineral water and sparkling water) and synthetic seawater were used as the aqueous environment. The commercial drinking waters were purchased at local supermarkets and the synthetic seawater was made via coursework Alternative B explanation.

2.1.   Synthetic Seawater

Three table-spoonfuls of ordinary culinary salt dissolved in 1.3 litres of tap water on jug jar were used to make the synthetic seawater.

Due to lack of available equipment to measure the chemical composition of artificial seawater, the information shown on the table 1 below was taken from the work carried out by Kester et al.

Table 1 – Synthetic seawater chemical composition

Synthetic Seawater	
Chemical Composition	g/Kg
Cl^-	19.35
Na^+	10.76
Mg^{+2}	1.297
Ca^{+2}	0.406
K^+	0.387
HCO_3^-	0.142
Br^-	0.066
Sr^{2+}	0.014
H_3BO_3	0.026
SO_4^{-2}	2.701
F^-	0.001
PH (measured on the experiment)	5

2.2.   Mineral Water

The mineral water used for this experiment was bought at KERO supermarket. The brand name is Agua Perla, see figure 2. The chemical composition is shown on the table 2 as below.

Table 2 – Mineral water chemical composition

Mineral Water	
Chemical Composition	mg/l
Ca^{+2}	3.5
Na^+	5.8
Cl^-	7
SO_4^{-2}	2
NO^{3-}	1
Mg^{+2}	2
SiO_2	15
HCO_3^-	41
PH (at source)	6.8

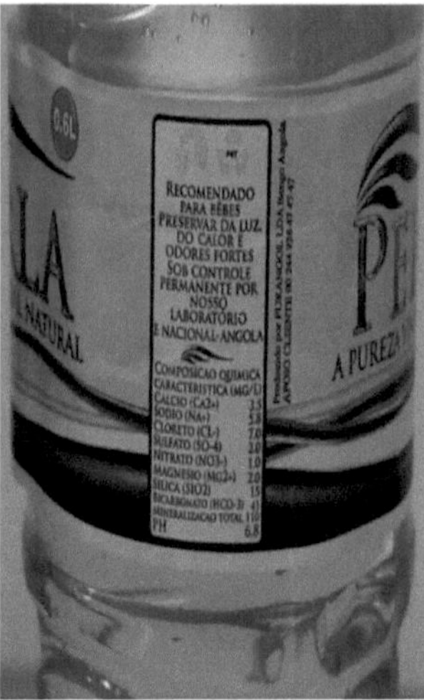

Figure 2 - Mineral water (Purchased at supermarket KERO)

## 2.3.	Sparkling water (CO_2 added)

The mineral water used for this experiment was bought at KERO supermarket. The brand name is FRIZE, see figure 3. The chemical composition is shown on the table 3 as below.

Table 3 – Sparkling water chemical composition

Sparkling Water Added CO_2	
Chemical Composition	mg/l
SiO_2	33
Cl^-	121
F	2
HCO_3^-	2062
Na^+	631
Mg^{+2}	31
Ca^{+2}	107
PH	6.15

Figure 3 - Sparkling water – CO_2 added (Purchased at supermarket KERO)

3. Visual Observations

Throughout the length period of the experiment the following visual observations were made and recorded as illustrated on the following sections.

3.1.   Synthetic seawater

On the first day of observations, no visual changes have been noticed in the first 30 minutes after immersing the iron nail on the synthetic seawater. The colour of the water and the nail remained the same. No strong discoloration could be seen. After 7.5 hours the iron nail was showing some evidence of corrosion attack, forming a light brown colour around the surface of the nail and also at the bottom of the jug jar. Within 2 days of observations the light brown colour around the nail, turned into dark brown, with some evidence of suspended rust particles around and at the bottom of the jug jar. After a week of observations the water and the nail turned all in dark brown colour. The nail was full covered with thick layers of rust (see figure 4). At the end of the experiment the dried nail, was fully dark brown in colour (see annex IV).

Figure 4 - Iron nail in synthetic seawater

3.2.   Mineral water

On the first day of observations, no visual changes have been noticed in after immersing the iron nail on the water. Like the synthetic seawater, the mineral water shown the same behaviour, after 7.5 hours the iron nail was showing some evidence of corrosion attack, forming a light brown colour around the surface of the nail and also the bottom of the bottle. Within 5 days of observations, the light brown colour around the nail, turned into dark brown covering all surface of the nail (see figure 5). Unlike in the synthetic seawater, only the nail turned dark brown, the mineral water was clear with some rust particles collected at the bottom of the bottle (see figure 5). At the end of experiment, the dried nail was brownish in colour, with the head and tip fully dark (see annex IV).

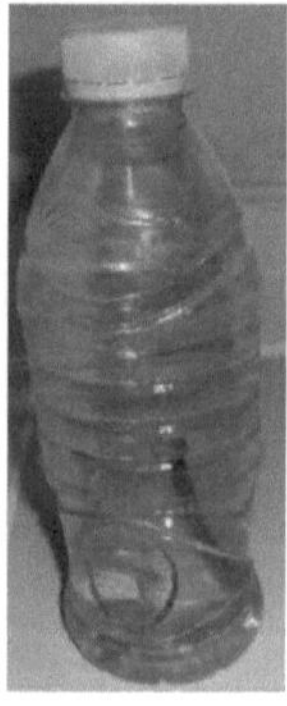

Figure 5 - Iron nail in mineral water

3.3.   Sparkling water (CO_2 added)

On the first day of observations, after immersing the iron nail, some gas bubbles started to surround the surface of the nail, whereas some of the gas bubbles moved very rapidly to the surface of water. After approximately 30 minutes, the gas bubbles around the nail, disappeared, remaining very few gas bubbles around it. No visual discolouration around the nail surface was noticed. This process remained constant throughout the experiment period. Towards the end of experiment, some thin layers in grey colours were seen at the top surface of the water. At the end of experiment, the dried nail was with very little traces of brown colour (see annex IV).

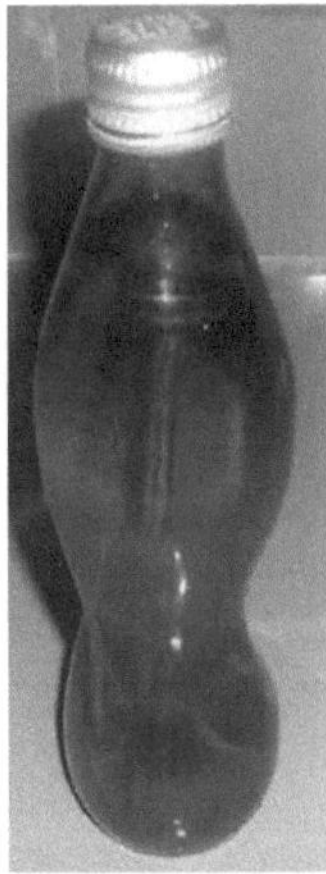

Figure 6 - Iron nail in sparkling water CO_2 added

4. Discussion

From what have been seen during the observation period in sparkling water, synthetic seawater and mineral water, combined with the ph data on the table available in annex I, the corrosion phenomena that takes place in the iron nails are explained in the following sub-sections.

4.1. Corrosion of Iron

The corrosion of iron to take place, the oxygen should dissolve in the water, reacting and causing iron corrosion and the formation of iron oxides. The following sections details further the chemical reactions.

4.1.1. Rust

Rusting is the corrosion of iron that occurs in alloy steel. The formation of reddish brown flakes which loosely adheres to the iron is known as rust. The iron nail is oxidized, and the oxidation half equation is expressed

$$2Fe_{[s]} \rightarrow 2Fe^{2+}_{[aq]} + 4e^-$$

Cathodic reduction of oxygen dissolved in water

$$O_{2[g]} + 2H_2O_{[l]} + 4e^- \rightarrow 4OH^-_{[aq]}$$

The overall equation

$$2Fe_{[s]} + 2H_2O_{[l]} + O_{2[g]} \rightarrow 2Fe^{2+}_{[aq]} + 4OH^-_{[aq]}$$

Fe^{2+} and OH^- ions migrate through the water by diffusion. When they meet they combine to produce the precipitate, iron (II) hydroxide $Fe(OH)_2$, which is further oxidized to iron (III) hydroxide $Fe(OH)_3$, and finally dehydrated to produce rust – $Fe_2O_3.nH_2O_{[s]}$.

The overall chemical equation for the formation of rust is

$$Iron + water + oxygen \rightarrow rust$$

$$4Fe_{[s]} + 6H_2O_{[l]} + 3O_{2[g]} \rightarrow 4Fe(OH)_{3[s]}$$

More details of chemical equations on rust formation, corrosion rate from metal loss and Pourbaix diagram, can be found in the annexes II and III.

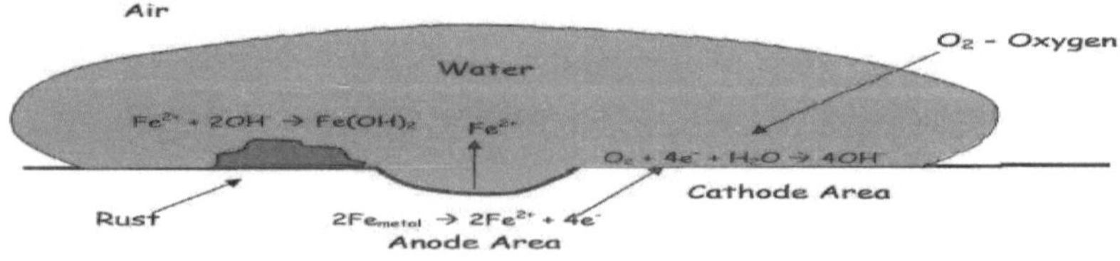

Figure 7 – Corrosion of iron in contact with aerated water (available from www.corrosionist.com/iron_corrosion.htm)

4.1.2.   Temperature Effect

In general corrosion rates increase with the increasing in temperature.
The temperature logged during the observation period was approximately
constant throughout that period, with an average value of 29 °C (refer
annex I). Therefore, it can be concluded that the effect of temperature on
the corrosion rate of iron nails (both in synthetic seawater and mineral
water) was minimal.

4.1.3.   pH Effect

The pH of the water alters the rate of corrosion as well as the mechanism
by which iron nails dissolves (Webster, 2010, pp.13). The graph of pH vs.
corrosion rate was plotted with values from log table available at annex I.
The pH values ranging from 4 to 10 shows the corrosion rate of iron,
being relatively independent of the pH of solution.

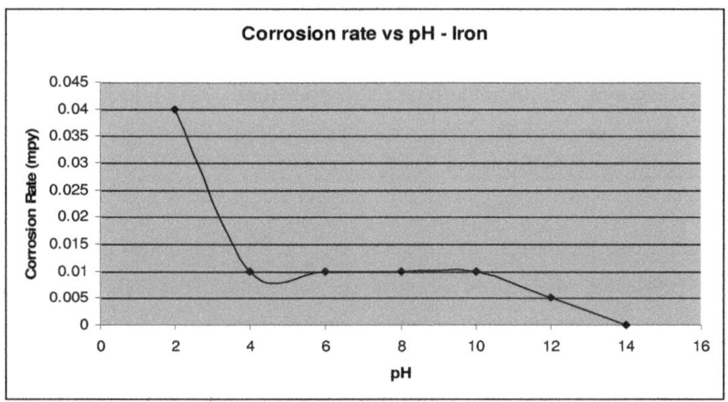

Graph 1 – Effect of pH on the corrosion rate of iron (adapted from
www.corrosionist.com)

The pH range (4 to 10) from the above graph clearly shows that the
corrosion rate is governed by the rate at which oxygen reacts with
absorbed atomic hydrogen, thereby depolarizing the surface and allowing
the reduction reaction to continue.

$$2H^+ + \tfrac{1}{2}O_2 \rightarrow H_2O - 2e^- \text{----- Depolarization}$$

4.2. CO_2 dissolved in water

Sparkling water contains CO_2 dissolved at low percentages $(0.2 - 1)$ %.
When CO_2 is dissolved in water, it forms a carbonic acid expressed as
follow:

$$CO_2 + H_2O \leftrightarrow H_2CO_3$$

Dissolved CO_2 in the form of H_2CO_3 may loose up two protons through the
acid equilibria:

$$H_2CO_{3[aq]} \leftrightarrow H^+_{[aq]} + HCO^-_{3[aq]}$$
$$HCO^-_{3[aq]} \leftrightarrow H^+_{[aq]} + CO_3^{2-}{}_{[aq]}$$

4.2.1 CO_2 corrosion mechanism

When iron (Fe) is immersed in CO_2 aqueous environment i.e. in sparkling
water, the overall electrochemical reaction is expressed as follow (Yang,
Brown and Nesic, 2013; Wang, 2009, pp 14):

$$Fe_{[s]} + CO_{2[g]} + H_2O_{[l]} \leftrightarrows FeCO_{3[s]} + H_{2[g]}$$

The solid iron carbonate precipitates when the concentration of Fe^{+2} and
CO_3^{2-} ions exceed the solubility limit (Sun and Nesic, 2006), forming a
surface scale. The solubility limit is exceeded at high pH, when
concentration CO_3^{2-} ions is high.

$$Fe^{+2} + CO_3^{2-} \leftrightarrows FeCO_{3[s]}$$

In these cases, the reaction shown above is not in equilibrium and there
is a net rate from left to right amounting to a formation of solid $FeCO3$.
This non equilibrium situation is termed supersaturation (Nesic, 2011, pp
236).

According to (Sun and Nesic, 2006) the iron carbonate scale slows down
the corrosion process by presenting a diffusion barrier for the species
involved in the corrosion process and by covering up a portion of the
steel surface and preventing the underlying steel from further dissolution.
Iron carbonate scale growth depends primarily on the precipitation
kinetics.

Most waters are saturated with respect to siderite $FeCO_3$. Siderite plays
an important role as an intermediate in the formation of protective layers
(Sander et al, 1996).

4.2.2.   pH effect

PH promotes the formation of $FeCO_3$ layers, and the kinetics of this process is highly temperature dependent (Nesic, 2011). Sparkling water contains bicarbonate HCO^-_3, and the typical pH values lies around 7 to 10 (refer annex I).

Wang 2009 (cited by Gray et al., 1989) mentioned that, an increase in pH will decrease the corrosion rate under non-film-forming conditions. It is considered to be non film- forming conditions when pH is less than 5. When pH is more than 6, it is relatively easy for $FeCO_3$ film to grow on the metal surface after reaching $FeCO_3$ supersaturation. The high pH level facilitates the passivation process, causing a sharp decline in corrosion rate.

4.2.3.   Temperature effect

Increased temperature accelerates all the processes involved in corrosion. At higher pH (> 5) the solubility of $FeCO_3$ is exceeded and forms a protective layer. Thus, the increasing in temperature accelerates the kinetics of precipitation of $FeCO_3$ and leads to formation of more protective layer (Nesic, 2011).

5. Conclusions

From the observations taken during the experiment period, it was concluded that iron nails tend to corrode faster when immersed in synthetic seawater as well as in mineral water. This rapid corrosion process involves the rapid formation of reddish brown flakes which adheres to the iron, known as rust. Whilst, the iron nail tend to corrode slower when immersed in sparkling water (CO_2 added). Thus, no visible trace of rust formation was seen. The solid iron carbonate scale played a key role in reducing the corrosion process in the CO_2 aqueous environment. The synthetic seawater and mineral water both had a low pH measured value i.e. 5, whilst the sparkling water had a high pH value i.e. 9.

6. References

AdiChemistry. "Chemistry of Carbonates & Bicarbonates (Part - 1)".
[Online]. Available at:
http://www.adichemistry.com/inorganic/p-block/group-
14/carbonates/carbonates-bicarbonates-1.html
Accessed on 15th June 2013.

Carbon Dioxide and Carbonic Acid. [Online]. Available at:
http://ion.chem.usu.edu/~sbialkow/Classes/3650/Carbonate/Carbonic%2
0Acid.html
Accessed on 2nd June 2013.

Chemguide (2012). Understandig Chemistry. [Online]. Available at:
http://www.chemguide.co.uk/mechmenu.html#top
Accessed on 2nd June 2013.

Chemical formula. "Rust". *Basic chemistry — writing chemical formula to
balancing chemical equations.* [Online]. Available at:
http://www.chemicalformula.org/rust
Accessed on 2nd June 2013.

Corrosionist. "Corrosion conversion rate". *The website of corrosion
protection and corrosion prevention.* [Online]. Available at:
http://www.corrosionist.com/corrosion_rate_conversion.htm
Accessed on 15th June 2013.

Gray, L. G. S., et al., (1989). "Mechanisms of carbon steel corrosion in
brines containing dissolved carbon dioxide at pH 4". *Corrosion 89*. Paper
No 464. NACE International. Houston, Texas.

Kester et al., (1967). " Preparation of artificial seawater". *Limnology and
Oceanography.* 12. pp. 176-179. [Online]. Available at
www.oslo.org/lo/toc/vol_12/issue_1/0176.pdf
Accessed on May 28th 2013

Nesic, S., (2011). " Carbon dioxide corrosion of mild steel". *Uhligs Corrosion Handbook.* 3rd ed. John Wiley & Sons.

Sander et al., (1996). " Iron corrosion in drinking water distribution systems − the effect of pH, calcium and hydrogen carbonate". *Corrosion Science.* 38(3), pp.443-455. Sciencedirect [Online]. Available at www.sciencedirect.com
Accessed on 11th June 2013

Wang. S., (2009). "Effect of oxygen on CO_2 corrosion of mild steel". Published Msc thesis. Russ College of Engineering and Technology. Ohio University. [Online]. Available at:
http://etd.ohiolink.edu/view.cgi?acc_num=ohiou1235976914
Accessed on 8th June 2013

Webster. S. K., (2010) "Effects of NaCl Concentration and Temperature on Corrosion: Understanding Material Interactions in Aqueous Environments". Published Msc thesis. Grand Valley State University. [Online]. Available at:
http://www.bumc.bu.edu/biomedforensic/files/2010/10/Katie-Webster-THESIS.pdf
Accessed on 11th June 2013

Yang. Y., Brown. B., Nesic. S., (2013) "Study of protective iron carbonate layer distribution in a CO2 corrosion environment". *Corrosion conference & expo.* Paper No 2708. NACE International. USA.

7. Annexes

Annex I

Log table

Date	Seawater			Mineral Water			Sparkling Water (CO2 added)		
	PH	Temp. °C	Obs	PH	Temp. °C	Obs	PH	Temp. °C	Obs
20/05/13	5	30		6	30		8	30	
22/05/13	4	29		5	29		9	29	
24/05/13	4	27		4	27		9	27	
30/05/13	7	29		6	29		9	29	
5/6/2013	5	28		5	28		8	28	
11/6/13	5	29		5	29		9	29	
17/06/13	5	29		5	29		9	29	
20/06/13	5	29		5	29		9	29	

Annex II

<u>Rust chemical equations</u>

$$2Fe_{[s]} + 2H_2O_{[l]} + O_{2[g]} \rightarrow 2Fe^{2+}_{[aq]} + 4OH^-_{[aq]} \text{-----------------------1}$$

$$2Fe^{2+}_{[aq]} + 2OH^-_{[aq]} \rightarrow Fe(OH)_{2[s]} \text{-------------------------------------2}$$

Iron (II) $Fe(OH)_{2[s]}$ is further oxidized to iron (III) $Fe(OH)_{3[s]}$. Iron (III) dehydrates to form $Fe_2O_3.nH_2O_{[s]}$ – Rust.

<u>Corrosion rate metal loss</u>

$$mm / y = 87.6*(W / DAT)$$

Where:

W = weight loss in milligrams

D = iron nail density in g/cm^3 (= 7.86 g/cm^3)

A = area of the sample bottles in cm^2

T = time of iron immersed in the sample in hrs (30 days = 720 hours)

Pourbaix diagram

The following Pourbaix diagram was used based on system iron – water at 25°C. The concentration of metallic species specified is 10^{-6} M. The electrochemical potentials are related to the standard hydrogen electrode SHE.

From the diagram in the pH range of 2 – 14 and a SHE of -0.5 V we see the non protective form of corroded iron i.e. the formation of rust Fe_2O_3. The iron will be in the stable region also known as immunity region if SHE is lowered to -0.6 V. Thus, the pH values measured in synthetic seawater and mineral water is around 5, and according with the Pourbaix diagram at SHE of (-0.6 to 0.8) V the iron starts to corrode to Fe^{2+}.

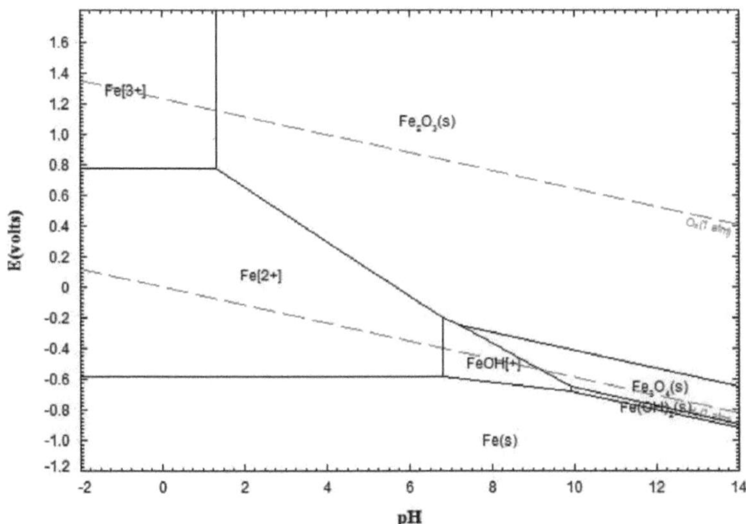

Pourbaix diagram (source: http://www.crct.polymtl.ca/ephweb.php!)

1. Dried nails after 25 days immersed in synthetic seawater

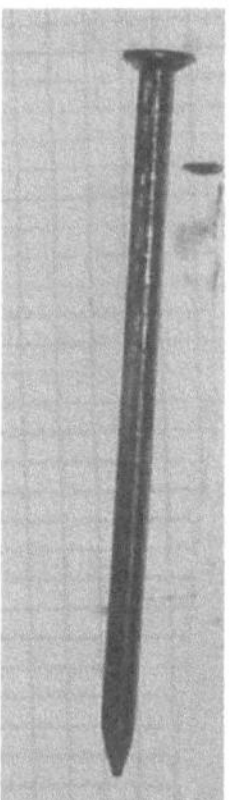

2. Dried nails after 25 days immersed in mineral water

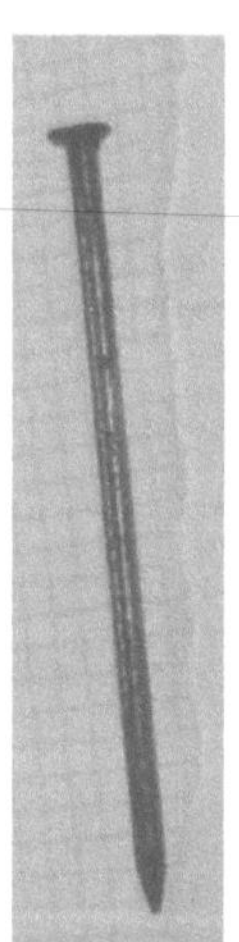
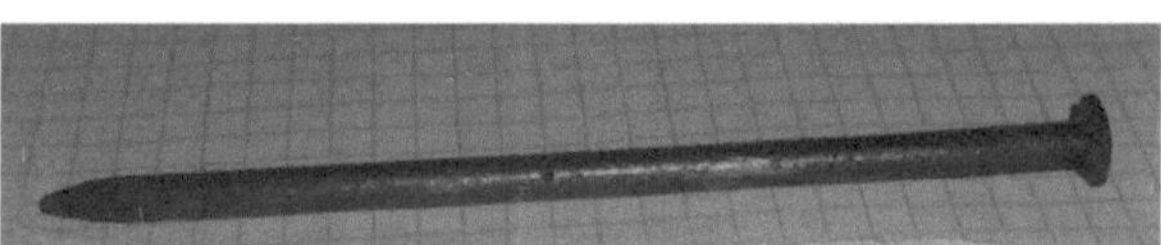

3. Dried nail after 25 days immersed in sparkling water